## THIS LOGBOOK BELONGS TO:

Name:

Class:                                    Date:

Science logbook number:                   of

☐  I acknowledge that this science logbook contains all my own work.

Your signature:

Teacher's signature:

Nelson VICscience Biology Logbook
1st Edition
ISBN 9780170452625

Publisher: Eleanor Gregory
Project editor: Robyn Beaver
Text design: Rina Gargano, Alba Design
Cover design: James Steer
Permissions researcher: Liz McShane
Production controller: Renee Tome
Typeset by: SPi Global

Any URLs contained in this publication were checked for currency during the production process. Note, however, that the publisher cannot vouch for the ongoing currency of URLs.

Acknowledgements
Extracts from the VCE Biology Study Design (2022–2026) and the VCE Biology Advice for Teachers are used by permission, © VCAA. VCE® is a registered trademark of the VCAA. The VCAA does not endorse or make any warranties regarding this study resource. Current VCE Study Designs, past VCE exams and related content can be accessed directly at www.vcaa.vic.edu.au.

For product information and technology assistance,
in Australia call 1300 790 853;
in New Zealand call 0800 449 725

For permission to use material from this text or product, please email
aust.permissions@cengage.com

ISBN 978 0 17 045262 5

Cengage Learning Australia
Level 7, 80 Dorcas Street
South Melbourne, Victoria Australia 3205

Cengage Learning New Zealand
Unit 4B Rosedale Office Park
331 Rosedale Road, Albany, North Shore 0632, NZ

For learning solutions, visit cengage.com.au

Printed in China by 1010 Printing International Limited.
1 2 3 4 5 6 7 24 23 22 21

# Contents

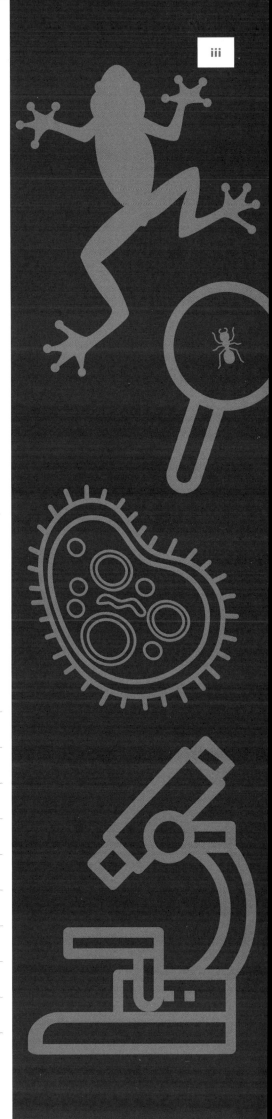

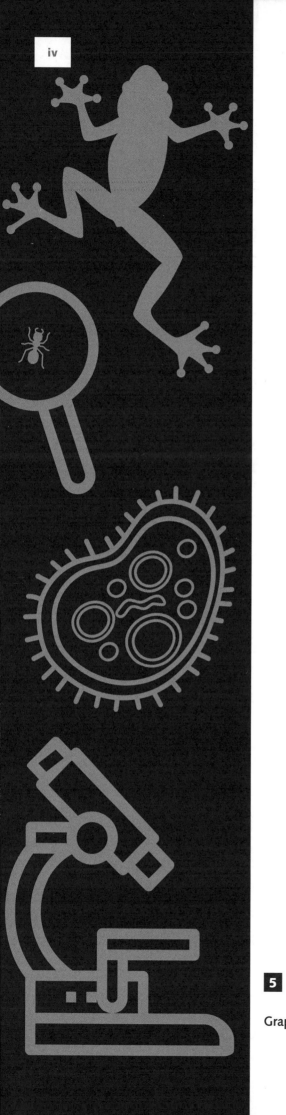

iv

9780170452625

# Why is your science logbook important?

## 1.1 Authentication

Investigation is the cornerstone of all sciences. Through performing investigations, you can experience what science is really like by finding things out through observation and experiments.

The Victorian Curriculum and Assessment Authority (VCAA) requires all VCE Biology students to undertake practical investigations as part of its VCE Biology course. Keeping a logbook to record information during scientific investigations and research is standard scientific practice. You will maintain a logbook that details all the investigations you will undertake as part of your VCE Biology course.

You will use your logbook to record all your practical and investigative work that involves the generation of primary data and/or the collection of secondary data. It is where you will write notes while performing investigations in class and it will help you when writing formal practical reports, revising and completing school-based assessments.

Your teacher will use your Biology logbook for **authentication** purposes. Authentication provides your teacher with evidence that you undertook each investigation and completed each entry in this logbook. This is important because the logbook is also used for **assessment** purposes and it is therefore essential that it is your own work that is being assessed. You are required to sign and date the bottom of each page of your logbook to verify that the work has been completed by you.

Your logbook will also help you to:
- contribute to class discussions
- report to the class on an investigation or activity
- respond to questions in a worksheet, workbook or problem-solving exercise
- write up an investigation as a formal report or a scientific poster.
  Keep in mind the following important tips for using your logbook.
- You must sign and date every logbook entry.
- All entries should be in chronological order.
- All entries must be your own work.
- Investigation partners, expert advice and assistance and secondary data sources must be acknowledged and/or referenced in your logbook.

## 1.2 How to use this book

This logbook provides a suggested approach to your investigative work and includes the VCAA VCE Biology key science skills table, an investigation scaffold with hints on how to approach your research to help you meet the VCAA requirements, and investigation templates that provide structure for your practical investigations.

The information you record in your logbook can be:
- ideas and planning notes for investigations
- investigations and/or student-designed activities
- observations made during or at the conclusion of demonstrations
- activities from the student book
- qualitative and/or quantitative results
- results of investigations.

Other helpful hints:

- The essential key science skills that you will learn and practise in your VCE Biology course are listed on pages 10–11. These key skills are examinable in the VCE Biology end-of-year exam.
- As you complete each investigation, add it to the contents list on pages iii–iv to easily locate your research.
- Complete the key terms list on pages 117–119 to build knowledge of important terms and their meanings.

## 1.3   Unit 1 Outcome 3: scientific investigation

In Unit 1, Area of Study 3, you will adapt or design, and then conduct, a scientific investigation. This investigation must be related to the function and/or regulation of cells or systems that you learn in Units 1 and 2. As part of this investigation, you are required to generate quantitative and/or qualitative primary data, organise and interpret that data and draw a conclusion from that data. This outcome provides you with the opportunity to demonstrate your understanding of the key science skills.

You may choose from several different presentation formats to communicate your findings, including a scientific poster, an article for publication in a scientific journal or magazine, a practical report, an oral presentation, a multimedia presentation or a visual presentation. Choose a format that will best communicate your findings.

## 1.4   Unit 4 Outcome 3: scientific poster

In Unit 4, Area of Study 3, you will demonstrate your science investigation and communication skills by presenting the findings of a student-designed scientific investigation from a topic from either Unit 3 or 4 or across Units 3 and 4. You will present your findings as a scientific poster. Your poster and logbook entries will both be assessed as part of Unit 4, Outcome 3.

The poster may be produced electronically or in hard-copy format and **should not exceed 600 words**.

The centre of the poster is a one-sentence summary of the key findings of the investigation that answers the research question. This occupies 20–25 per cent of the space and is surrounded by the key components of scientific reports.

Refer to the following scientific poster requirements (Table 1) and template (Figure 1) when reporting on your investigation.

**Table 1** Sections of a scientific poster

| Poster section | Content |
|---|---|
| Title | The research question that you are investigating |
| Introduction | A brief explanation or reason for undertaking the investigation, including:<br>• a clear aim<br>• a hypothesis and/or prediction<br>• relevant background biological concepts |
| Methodology and methods | A brief outline of the selected methodology used to address the investigation question. It should be authenticated by logbook entries.<br>A summary of data generation method/s and data analysis method/s. |
| Results | Presentation of generated data/evidence in appropriate formats to illustrate trends, patterns and/or relationships. Formats may include a results table and/or a graph or pie chart, depending on the type of data collected. |
| Discussion | • Interpretation and evaluation of analysed primary data.<br>• Identification of limitations in the data and methods, including the identification of outliers, and suggested improvements.<br>• Cross-referencing of results to relevant biological concepts.<br>• Linking of results to the investigation question and to the aim to explain whether or not the investigation data and findings support the hypothesis.<br>• Implications of the investigation and/or suggestions as to further investigations that may be undertaken. |

| Poster section | Content |
|---|---|
| Conclusion | • Conclusion that provides a response to the investigation question.<br>• Identification of the extent to which the analysis has answered the investigation question, with no new information being introduced. |
| References and acknowledgements | Referencing and acknowledgement of all quotations and sourced content relevant to the investigation. (Note: Not part of poster word count.) |

Source: adapted from the *VCE Biology Study Design* (2022–2026) pp. 11–12; © VCAA, by permission

It is essential to record all elements of your investigation in your logbook. This includes planning, including identification and management of relevant risks; recording of raw data; and preliminary analysis and evaluation of your results, including identification of outliers and their subsequent treatment. The design, analysis and findings of your investigation will be presented as a scientific poster using the format shown in Figure 1.

| Title |
|---|
| Student name |

| Introduction | | Discussion |
|---|---|---|
| Methodology and methods | Communication statement reporting the key finding of the investigation as a one-sentence summary (20–25% of poster space) | |
| Results | | Conclusion |
| References and acknowledgements | | |

Source: adapted from the *VCE Biology Study Design* (2022–2026) p. 11; © VCAA, by permission

**Figure 1** Poster template

The VCAA provides extra information on effective scientific poster communication. You can find the most up-to-date advice on the VCAA website (www.vcaa.vic.edu.au).

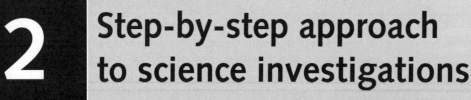

# 2 | Step-by-step approach to science investigations

The investigation templates in this book follow the standard format described below.

## 1 Title

This should be a clear, descriptive title that outlines the investigation being written. The title should be written as a research question. For example:

*Is the activity of the enzyme trypsin affected by pH?*

## 2 Introduction, aim and hypothesis

Record the specific information that you find during the research phase of your investigation.

For example, in this case of enzyme activity and pH, the following questions could be addressed.

*What is meant by pH?*

*Can you describe what an enzyme is?*

*Can you explain how enzymes work?*

*Can you describe what is meant by an 'active site' and how its shape affects an enzyme's activity?*

Remember to record all **references** of your sources (using the correct format you have chosen) and recognise any contributors by including them in the **acknowledgements section.**

### Aim

State the aim of the experiment:

*To determine if the activity of the enzyme trypsin is affected by pH*

### Hypothesis

The hypothesis is a prediction of what you think will happen, based on your research and scientific knowledge. The hypothesis:

*   includes the independent variable and the dependent variable, including any applicable units or how it is to be measured.
*   must be both testable and include a prediction.
*   is written in the third person.

The VCAA does not mandate any one way of writing a hypothesis. A common format for writing a hypothesis is the 'If ... then ...' format. There are other suitable formats that you can also use to write a hypothesis.

| Step | Description | Example |
|------|-------------|---------|
| 1 | Ask a research question of interest. | Is the activity of the enzyme trypsin affected by pH? |
| 2 | Identify the independent variable (IV). This is the variable for which quantities are manipulated (selected or changed) by the experimenter and assumed to have a direct effect on the dependent variable. Independent variables are plotted on the x (horizontal) axis of graphs. | pH |
| 3 | Identify the dependent variable (DV). This is the variable the experimenter measures, after selecting the independent variable that is assumed to affect the dependent variable. Dependent variables are plotted on the y (vertical) axis of graphs. | Activity of the enzyme |
| 4 | Construct a hypothesis. | **If** the activity of the enzyme (DV) is affected by pH **then** the activity of the enzyme will decrease when the pH is changed from a pH of 7 (IV). |

Here are some more examples of hypotheses written using the **If... then...** format:
- **If** the height of tomato seedlings (DV) is related to the amount of water they receive **then** tomato plants receiving more water will grow higher than those receiving less water (IV).
- **If** the number of mice (DV) in an area is related to the amount of food available **then** there will be more mice in an area where there is plentiful food (IV).
- **If** bacterial growth increases with temperature **then** bacterial growth will increase as temperature increases (IV).

In summary: **If** the DV is related to the IV **then** the DV will be affected in some way when the IV is changed.

## 3 Methodology and methods

### Methodology

The methodology refers to the broad framework of the approach taken in the investigation to test your hypothesis. See page 117 for a list of the methodologies from which you can choose to test your research question or hypothesis.

### Variables

Identify each of the variables in the experiment, what type they are, and how they are controlled or measured. You identified the independent variable (IV) and dependent variable (DV) during the development of your hypothesis. **Controlled variables** should also be listed and specifically defined. These are the variables that need to be kept the same for the experiment to be fair.

A **control** should also be set up. An example of a control in this case would be a test tube with substrate and enzyme at pH 7 (neutral). No acidic/basic solution should be added to this test tube. This is used to compare directly with the independent variable to ensure that the dependent variable (the activity of the enzyme, as measured) is due to the independent variable only (the change in pH) and does not also happen when the IV is not present.

Note, questions about the control often appear in VCAA VCE Biology exams.

**Table 3** Examples of variables for the above enzyme hypothesis

| Variable | Type |
|----------|------|
| pH | Independent |
| Activity of the enzyme | Dependent |
| Concentration of trypsin | Controlled variable |
| Volume of trypsin | Controlled variable |
| Concentration of substrate | Controlled variable |
| Volume of substrate | Controlled variable |
| Size of test tubes | Controlled variable |
| Temperature of set up | Controlled variable |
| Control | Test tube containing substrate and enzyme at pH 7 |

## Materials

List all the equipment and materials that you need to perform and carry out your investigation. Include, where relevant, the number of items, volume and concentration of liquids and amounts of materials required.

**Example of a materials list:**

- 1% trypsin solution
- 3% solution of skim milk powder
- Range of solutions from pH 4 to pH 11
- 16 test tubes
- Bungs or corks for test tubes
- 2 test-tube racks
- Stopwatch
- Marker pen
- Plastic pipettes
- Lab coats
- Safety glasses
- Disposable gloves
- Timer or stopwatch

## Risk assessment

- List and describe any risks or hazards in your experiment. This can include physical, biological and chemical hazards.
- You need to refer to relevant **Safety Data Sheets (SDS)** if you are using hazardous chemicals as part of your method.
- Familiarise yourself with Personal Protective Equipment (PPE).

**Table 4** Risk assessment example

| What are the risks in this investigation? | How can you manage these risks to stay safe? |
|---|---|
| Trypsin can cause allergic reactions in sensitive people. | Always wear appropriate PPE, including eye protection and gloves. Wash skin immediately if contact occurs. |
| Trypsin can be irritating to the skin and eyes on contact. | Always wear appropriate PPE, including eye protection and gloves. Wash skin immediately if contact occurs. |
| Disposable gloves can pose an allergy risk. | Use a glove type that removes allergy risk and is suitable for the chemicals being used. |
| Solutions of low and high pH can be irritating to the skin and eyes on contact. | Always wear appropriate PPE, including eye protection and gloves. Wash skin immediately if contact occurs. |

## Method

The method shows the 'steps' required to carry out your investigation. If you make changes to your method after you have written it, make sure you record the changes. The instructions should be written in a numbered format so that someone else can easily follow your method to reproduce your investigation and achieve similar results. Remember, data must be reproducible. Include a diagram to show how to set up the equipment (if relevant). Trial 1 below shows the method for the example experiment.

Remember, that data must be reproducible.

### Trial 1

1. Collect 8 test tubes. Mark each with an 'X' on the glass halfway down the tube.
2. Using a pipette, add 10 mL of the milk powder solution to each of the 8 test tubes.
3. Collect the other 8 test tubes, add 3 mL of solution of pH 4 to the first test tube, pH 5 to next and so on. Then add 3 mL of trypsin solution to each.
4. Pour the trypsin and acidic/basic solution from one test tube into the milk powder solution into each of the other test tubes.
5. To mix thoroughly, place a cork in each test tube and invert approximately 5 times.
6. Place each test tube into the test-tube rack.
7. Record the time it takes for the milk solution to become as clear as 'Test 1' and for the 'X' to be visible through the solution.

**Figure 2** Test tube showing X becoming visible through the milky solution

### Trial 2
**8** Repeat this process (steps 1–7).
**9** Calculate the average reaction time and record the result.

## 4 Results
### Results table
Present your results in a table. Make sure:
- the table has a title (What is it showing?)
- each column has a heading with appropriate units
- all data is presented with a consistent number of decimal places
- all of your raw data is shown and the **average** or **mean** is calculated.

**Table 5** Results table example

| pH | Trial 1 (sec) | Trial 2 (sec) | Mean time to clear (sec) |
|---|---|---|---|
| 4 | | | |
| 5 | | | |
| 6 | | | |
| 7 | | | |
| 8 | | | |
| 9 | | | |
| 10 | | | |
| 11 | | | |

Include all of your comments in your logbook and ensure you record any qualitative changes or observations. The best way to display quantitative results is to draw a chart or graph. The graph or chart should:

- have a title
- use as much of the graph paper as possible (Work out the correct scale to use on the axes.)
- have a labelled *x*-axis (with units) (independent variable)
- have a labelled *y*-axis (with units) (dependent variable).
  When you plot your data:
- use a pencil to draw your axes and plot the points but use a pen for labels
- mark the points using crosses and not dots
- join the points using a ruler or draw a line of best fit.

# 5 Discussion

In the discussion section, you will use the evidence that you have produced from your investigation to construct a scientific argument about how well your investigation answered your research question and achieved the intended aim(s). This is probably the most difficult part of the report to write. The following is a suggestion for structuring the discussion section.

**Step 1**: Make clear statements about any trends you identified in the data.

*At the extremes of the pH (4 and 11), the activity of the trypsin was very slow. The activity of the trypsin increased as the pH approached neutral (7) and was most active at pH 7.*

**Step 2**: Explain the science behind the observed trends. It is important that you refer to your background research to explain the results. Remember to include biological terminology and scientific concepts.

*The optimal pH for the activity of the enzyme trypsin is 7. As the pH moves towards more acidic or more basic, the activity of the enzyme begins to decrease. Towards the lower (4) and higher (11) ends of the pH range, the activity almost stopped. Trypsin is made from protein and the shape of the protein, particularly the active site, is important in enabling this reaction to occur. High and low pH solutions denature, or change the shape of the active site, causing the reactants to no longer fit and the reaction to slow or stop.*

**Step 3**: Evaluate your data in terms of accuracy and precision and state if you think the results are valid. Have you identified any:

- human errors – did you make any mistakes recording the results or calculating the mean?
- outliers – are there any results that differ markedly from the other results?

**Step 4**: Suggest how you can improve the investigation to eliminate any errors.

*Another trial could be added to repeat the process to eliminate any human errors and outliers.*

**Step 5**: Link your results to the research question and the aim to explain if the data and findings support the hypothesis.

*This investigation was to discover if the activity of the enzyme trypsin is affected by pH. The results show that at the lower and higher levels of the pH scale, the activity of trypsin decreased. At pH 4 it took 238 seconds for the mixture to clear and at pH 11 it took 253 seconds for the mixture to clear. As the pH approached neutral, the activity of the trypsin increased. At pH 6 it took 45 seconds for the mixture to clear and at pH 8 it took 52 seconds for the mixture to clear. At pH 7, the mixture cleared in 21 seconds.*

*These results support the hypothesis that the activity of trypsin is affected by the pH of the solution.*

**Step 6**: Suggest how you could extend or further investigate the topic.

*The pH could be extended from pH 1 up to pH 14 to provide further evidence of the effect of pH on trypsin activity.*

# 6 Conclusion

This short section allows you to draw conclusions based on the evidence you have gathered during your study. It is a brief summary of the results and their implications. It should provide a response to your research question and directly address the hypothesis you proposed in your introduction. The conclusion should also state the extent to which the analysis answered the research question, without introducing new information. A conclusion is only a few sentences long.

*This investigation showed that the enzyme trypsin is most active at pH 7. As the pH moved towards the higher and lower ends of the range, the activity of the enzyme decreased. These results support the hypothesis, but further investigation into the action of trypsin at the far ends of the pH range could be explored.*

# 7 References and acknowledgements

At the end of well-conducted academic research you will find a list of references and contributors. Acknowledging all sources of information used in your research demonstrates your understanding of integrity in relation to the collection and reporting of data and an ethical approach to your investigation. This is also a VCAA requirement.

## References

This is a detailed list of the secondary sources that you have used in your investigation. They may be textbooks, websites, or journal or newspaper articles that you have quoted or paraphrased in your report, or images, diagrams or tables that you found on the internet or in books. It is important that every source used is included in the references list and that the references follow a consistent format. Two of the best-known referencing formats are American Psychological Association (APA) and Harvard referencing style, both of which are author–date systems. Check with your teacher before selecting a method.

In 'author–date' style, the same referencing format applies to books, journal articles, internet documents and other material. In general, the sequence to follow is:

1   Author or authors. The surname is followed by first initials. Note that the author may be an organisation.
2   Year of publication of the book or article, in brackets. If there is no date, use the abbreviation (n.d) for (no date).
3   Book title (in italics) or article title (in quotation marks).
4   Journal title (in italics)
5   Volume and issue number of journal, followed by page range of article
6   Publisher of book
7   Date accessed (for digital content).

**Example of a book reference:**
Chidrawi G, Bradstock S, Robson M and Thrum E (2019) *Biology in Focus*, 2nd edn, Cengage Learning Australia.
**Example of a journal article reference:**
King K. D. (2020) 'Patterns in genetic mutation', *International Journal of Genetics*, 22(9): 135–212.
**To reference a website, include the URL and the date accessed:**
http://www.carrotmuseum.co.uk, accessed 2 September 2020.
**To reference online images:**
Figure 2: photograph of modern carrot varieties (n.d.), http://www.carrotmuseum.co.uk/history7.html, accessed 20 September 2020.

## Acknowledgements

This is your opportunity to thank or acknowledge people who have helped in the planning or execution of your investigation. They may have given you general advice, or supplied materials or other assistance, such as giving you access to specialist equipment. Be specific about the contribution each person has made and ensure that the person's job title and organisation they work for are included. Always double check the spelling of your contributors' names.

**Example of an acknowledgement:**

> *Thanks to Julie Brooking, Head of Science, Palm View Secondary College, for her creative criticism of my investigation plan and her suggestions for improvement. Thanks also to Ron Brown, Laboratory manager, Biology Mart Pty Ltd, for providing the trypsin solution for my experiments.*

Note: the reference and acknowledgement section of the poster does not contribute to the poster's 600 word count.

# 3 Key science skills

During your VCE Biology studies, you will undertake many scientific investigations where you will have the opportunity to use and apply the key science skills listed below. As you continue your study of Biology, you will reinforce and build on these skills. These skills apply to all four units of Biology and are examinable in tests and exams.

| Key science skill | VCE Biology Units 1–4 |
|---|---|
| Develop aims and questions, formulate hypotheses and make predictions | • identify, research and construct aims and questions for investigation<br>• identify independent, dependent and controlled variables in controlled experiments<br>• formulate hypotheses to focus investigation<br>• predict possible outcomes |
| Plan and conduct investigations | • determine appropriate investigation methodology: case study; classification and identification; controlled experiment; correlational study; fieldwork; literature review; modelling; product, process or system development; simulation<br>• design and conduct investigations; select and use methods appropriate to the investigation, including consideration of sampling technique and size, equipment and procedures, taking into account potential sources of error and uncertainty; determine the type and amount of qualitative and/or quantitative data to be generated or collated<br>• work independently and collaboratively as appropriate and within identified research constraints, adapting or extending processes as required and recording such modifications |
| Comply with safety and ethical guidelines | • demonstrate safe laboratory practices when planning and conducting investigations by using risk assessments that are informed by safety data sheets (SDS), and accounting for risks<br>• apply relevant occupational health and safety guidelines while undertaking practical investigations<br>• demonstrate ethical conduct when undertaking and reporting investigations |
| Generate, collate and record data | • systematically generate and record primary data, and collate secondary data, appropriate to the investigation, including use of databases and reputable online data sources<br>• record and summarise both qualitative and quantitative data, including use of a logbook as an authentication of generated or collated data<br>• organise and present data in useful and meaningful ways, including schematic diagrams, flow charts, tables, bar charts and line graphs<br>• plot graphs involving two variables that show linear and non-linear relationships |

| Key science skill | VCE Biology Units 1–4 |
|---|---|
| Analyse and evaluate data and investigation methods | • process quantitative data using appropriate mathematical relationships and units, including calculations of ratios, percentages, percentage change and mean<br>• identify and analyse experimental data qualitatively, handling where appropriate concepts of: accuracy, precision, repeatability, reproducibility and validity of measurements; errors (random and systematic); and certainty in data, including effects of sample size in obtaining reliable data<br>• identify outliers, and contradictory or provisional data<br>• repeat experiments to ensure findings are robust<br>• evaluate investigation methods and possible sources of personal errors/mistakes or bias, and suggest improvements to increase accuracy and precision, and to reduce the likelihood of errors |
| Construct evidence-based arguments and draw conclusions | • distinguish between opinion, anecdote and evidence, and scientific and non-scientific ideas<br>• evaluate data to determine the degree to which the evidence supports the aim of the investigation, and make recommendations, as appropriate, for modifying or extending the investigation<br>• evaluate data to determine the degree to which the evidence supports or refutes the initial prediction or hypothesis<br>• use reasoning to construct scientific arguments, and to draw and justify conclusions consistent with the evidence and relevant to the question under investigation<br>• identify, describe and explain the limitations of conclusions, including identification of further evidence required<br>• discuss the implications of research findings and proposals |
| Analyse, evaluate and communicate scientific ideas | • use appropriate biological terminology, representations and conventions, including standard abbreviations, graphing conventions and units of measurement<br>• discuss relevant biological information, ideas, concepts, theories and models and the connections between them<br>• analyse and explain how models and theories are used to organise and understand observed phenomena and concepts related to biology, identifying limitations of selected models/theories<br>• critically evaluate and interpret a range of scientific and media texts (including journal articles, mass media communications and opinions in the public domain), processes, claims and conclusions related to biology by considering the quality of available evidence<br>• analyse and evaluate bioethical issues using relevant approaches to bioethics and ethical concepts, including the influence of social, economic, legal and political factors relevant to the selected issue<br>• use clear, coherent and concise expression to communicate to specific audiences and for specific purposes in appropriate scientific genres, including scientific reports and posters<br>• acknowledge sources of information and assistance, and use standard scientific referencing conventions |

Source: *VCE Biology Study Design* (2022–2026) pp. 7–9; © VCAA, reproduced by permission

# 4 My investigations

## INVESTIGATION

**Investigation number**

**Title**

> The title is the research question under investigation.

**Introduction**

> Reason for undertaking the investigation, any relevant background research and biological concepts that need to be explained.

**Aim**

> What are you trying to find out?

# Hypothesis

Use the format: **If** (the DV is related to the IV) **then** (the DV will be affected when the IV is changed).

# Methodology

Variables that could be changed.

The variable that you will change (IV).

The variable that you will measure (DV).

The variables that you will keep the same (controlled variables).

The **control** that will be set up.

# Materials

The equipment and materials that you will use.

## Method

A numbered step-by-step
description of how you plan
to test the hypothesis.

You can use the blank pages
in the back of this book to
draw, cut and paste your
experimental set up here.

Risk assessment
Do you need to check any Safety Data Sheets?

| What are the risks in doing this investigation? | How can you manage these risks to stay safe? |
|---|---|
| | |

## Results

Draw a results table and include the units in the
headings. Include a title.

| Independent variable (units) | Dependent variable (units) | | | |
|---|---|---|---|---|
| | Test 1 | Test 2 | Test 3 | Mean |
| | | | | |

Depending on the data, use
graph paper from the back of
this book to plot a graph or
bar chart. Include a title, label
the axes and include units.
Cut out your graph and glue
it here.

Remember, the dependent
variable is on the y-axis
(vertical) and the independent
variable on the x-axis
(horizontal).

Name:                     Your signature:                     Date:

Partner:                  Partner's signature:                Teacher's signature:

9780170452625
9780170452625

# Discussion

Make a clear statement about any trends identified in the data.

Explain the science behind the observed results.

Evaluate the data in terms of accuracy, precision, validity and identify any outliers and errors.

Suggest how the investigation could be improved to eliminate errors.

Link your results to the research question and aim.

Do your results support or refute your hypothesis?

Suggest how you could extend or further investigate this topic.

| Name: | Your signature: | Date: |
|---|---|---|
| Partner: | Partner's signature: | Teacher's signature: |

9780170452625
9780170452625

## Conclusion

Draw your conclusion based on the evidence you have gathered in this investigation.

Respond to your research question and hypothesis.

To what extent did your analysis answer the research question?

Do not introduce any new information.

## References and acknowledgements

Have you included ALL your sources of information?

Have you used a consistent and correct referencing convention?

See page 9 for a reminder of referencing methods

Name:      Your signature:      Date:

Partner:      Partner's signature:      Teacher's signature:

9780170452625
9780170452625

## INVESTIGATION

**Investigation number**

**Title**

**Introduction**

**Aim**

**Hypothesis**

Name: Your signature: Date:
Partner: Partner's signature: Teacher's signature:

## Methodology

## Materials

## Method

**Results**

| Name: | Your signature: | Date: |
|---|---|---|
| Partner: | Partner's signature: | Teacher's signature: |

## Discussion

## Conclusion

## References and acknowledgements

| Name: | Your signature: | Date: |
|---|---|---|
| Partner: | Partner's signature: | Teacher's signature: |

## INVESTIGATION

**Investigation number**

**Title**

**Introduction**

**Aim**

**Hypothesis**

## Methodology

## Materials

## Method

Name:                          Your signature:                          Date:

Partner:                       Partner's signature:                     Teacher's signature:

**Results**

## Discussion

Name:      Your signature:      Date:

Partner:      Partner's signature:      Teacher's signature:

## Conclusion

## References and acknowledgements

## INVESTIGATION

**Investigation number**

**Title**

**Introduction**

**Aim**

**Hypothesis**

| Name: | Your signature: | Date: |
|---|---|---|
| Partner: | Partner's signature: | Teacher's signature: |

# Methodology

## Materials

## Method

## Results

## Discussion

## Conclusion

## References and acknowledgements

Name:                          Your signature:                          Date:

Partner:                    Partner's signature:                    Teacher's signature:

## INVESTIGATION

**Investigation number**

**Title**

**Introduction**

**Aim**

**Hypothesis**

## Methodology

## Materials

## Method

| Name: | Your signature: | Date: |
| --- | --- | --- |
| Partner: | Partner's signature: | Teacher's signature: |

**Results**

## Discussion

Name:                    Your signature:                    Date:

## Conclusion

## References and acknowledgements

## INVESTIGATION

**Investigation number**

**Title**

**Introduction**

**Aim**

**Hypothesis**

| Name: | Your signature: | Date: |
|---|---|---|
| Partner: | Partner's signature: | Teacher's signature: |

## Methodology

## Materials

## Method

Name:

Your signature:

Date:

Partner:

Partner's signature:

Teacher's signature:

9780170452625

**Results**

Name:   Your signature:   Date:

Partner:   Partner's signature:   Teacher's signature:

## Discussion

Name:

Your signature:

Date:

Partner:

Partner's signature:

Teacher's signature:

9780170452625

## Conclusion

## References and acknowledgements

Name: _____ Your signature: _____ Date: _____

Partner: _____ Partner's signature: _____ Teacher's signature: _____

## INVESTIGATION

**Investigation number**

**Title**

**Introduction**

**Aim**

**Hypothesis**

## Methodology

## Materials

## Method

Name:     Your signature:     Date:

Partner:     Partner's signature:     Teacher's signature:

**Results**

**Discussion**

Name:                          Your signature:                          Date:

Partner:                       Partner's signature:                     Teacher's signature:

## Conclusion

## References and acknowledgements

## INVESTIGATION

**Investigation number**

**Title**

**Introduction**

**Aim**

**Hypothesis**

Name:                          Your signature:                          Date:

Partner:                       Partner's signature:                     Teacher's signature:

## Methodology

## Materials

## Method

Name:

Your signature:

Date:

Partner:

Partner's signature:

Teacher's signature:

9780170452625

**Results**

## Discussion

## Conclusion

## References and acknowledgements

| Name: | Your signature: | Date: |
|---|---|---|
| Partner: | Partner's signature: | Teacher's signature: |

## INVESTIGATION

**Investigation number**

**Title**

**Introduction**

**Aim**

**Hypothesis**

## Methodology

## Materials

## Method

| Name: | Your signature: | Date: |
|---|---|---|
| Partner: | Partner's signature: | Teacher's signature: |

**Results**

Name:
Your signature:
Date:
Partner:
Partner's signature:
Teacher's signature:
9780170452625

## Discussion

| Name: | Your signature: | Date: |
|---|---|---|
| Partner: | Partner's signature: | Teacher's signature: |

## Conclusion

## References and acknowledgements

## INVESTIGATION

**Investigation number**

**Title**

**Introduction**

**Aim**

**Hypothesis**

Name:                     Your signature:                     Date:

Partner:                  Partner's signature:                Teacher's signature:

**Methodology**

**Materials**

**Method**

## Results

Name:  Your signature:  Date:
Partner:  Partner's signature:  Teacher's signature:

## Discussion

## Conclusion

## References and acknowledgements

Name:      Your signature:      Date:

Partner:      Partner's signature:      Teacher's signature:

## INVESTIGATION

**Investigation number**

**Title**

**Introduction**

**Aim**

**Hypothesis**

## Methodology

## Materials

## Method

Name:

Your signature:

Date:

Partner:

Partner's signature:

Teacher's signature:

**Results**

**Discussion**

| Name: | Your signature: | Date: |
| --- | --- | --- |
| Partner: | Partner's signature: | Teacher's signature: |

## Conclusion

## References and acknowledgements

## INVESTIGATION

**Investigation number**

**Title**

**Introduction**

**Aim**

**Hypothesis**

## Methodology

## Materials

## Method

**Results**

Name:                    Your signature:                    Date:

Partner:                 Partner's signature:              Teacher's signature:

## Discussion

## Conclusion

## References and acknowledgements

Name:     Your signature:     Date:

Partner:     Partner's signature:     Teacher's signature:

## INVESTIGATION

**Investigation number**

**Title**

**Introduction**

**Aim**

**Hypothesis**

## Methodology

## Materials

## Method

| Name: | Your signature: | Date: |
|---|---|---|
| Partner: | Partner's signature: | Teacher's signature: |

**Results**

Name:
Partner:
Your signature:
Partner's signature:
Date:
Teacher's signature:
9780170452625

## Discussion

| Name: | Your signature: | Date: |
|---|---|---|
| Partner: | Partner's signature: | Teacher's signature: |

## Conclusion

## References and acknowledgements

Name:

Your signature:

Date:

Partner:

Partner's signature:

Teacher's signature:

9780170452625

## INVESTIGATION

**Investigation number**

**Title**

**Introduction**

**Aim**

**Hypothesis**

| Name: | Your signature: | Date: |
|---|---|---|
| Partner: | Partner's signature: | Teacher's signature: |

## Methodology

## Materials

## Method

**Results**

Name:      Your signature:      Date:

Partner:      Partner's signature:      Teacher's signature:

## Discussion

**Conclusion**

**References and acknowledgements**

Name: Your signature: Date:

Partner: Partner's signature: Teacher's signature:

## INVESTIGATION

**Investigation number**

**Title**

**Introduction**

**Aim**

**Hypothesis**

## Methodology

## Materials

## Method

| Name: | Your signature: | Date: |
|---|---|---|
| Partner: | Partner's signature: | Teacher's signature: |

**Results**

Name:

Your signature:

Date:

Partner:

Partner's signature:

Teacher's signature:

9780170452625

## Discussion

| Name: | Your signature: | Date: |
|---|---|---|
| Partner: | Partner's signature: | Teacher's signature: |

## Conclusion

## References and acknowledgements

## INVESTIGATION

**Investigation number**

**Title**

**Introduction**

**Aim**

**Hypothesis**

| Name: | Your signature: | Date: |
|-------|-----------------|-------|
| Partner: | Partner's signature: | Teacher's signature: |

## Methodology

## Materials

## Method

**Results**

| Name: | Your signature: | Date: |
| --- | --- | --- |
| Partner: | Partner's signature: | Teacher's signature: |

## Discussion

## Conclusion

_____

_____

_____

_____

_____

_____

_____

_____

_____

_____

_____

_____

_____

## References and acknowledgements

_____

_____

_____

_____

_____

_____

_____

_____

_____

_____

_____

_____

_____

Name: _____ Your signature: _____ Date: _____

Partner: _____ Partner's signature: _____ Teacher's signature: _____

## INVESTIGATION

**Investigation number**

**Title**

**Introduction**

**Aim**

**Hypothesis**

# Methodology

# Materials

# Method

Name:      Your signature:      Date:

Partner:      Partner's signature:      Teacher's signature:

**Results**

## Discussion

## Conclusion

## References and acknowledgements

## INVESTIGATION

**Investigation number**

**Title**

**Introduction**

**Aim**

**Hypothesis**

Name:                     Your signature:                     Date:

## Methodology

## Materials

## Method

Name: _____  Your signature: _____  Date: _____

Partner: _____  Partner's signature: _____  Teacher's signature: _____

9780170452625

**Results**

Name:                          Your signature:                          Date:

Partner:                       Partner's signature:                     Teacher's signature:

## Discussion

Name:

Your signature:

Date:

Partner:

Partner's signature:

Teacher's signature:

9780170452625

## Conclusion

## References and acknowledgements

Name:                          Your signature:                          Date:

Partner:                       Partner's signature:                     Teacher's signature:

## INVESTIGATION

**Investigation number**

**Title**

**Introduction**

**Aim**

**Hypothesis**

## Methodology

## Materials

## Method

Name:                Your signature:                Date:

Partner:                Partner's signature:                Teacher's signature:

**Results**

## Discussion

Name:     Your signature:     Date:

Partner:     Partner's signature:     Teacher's signature:

# Conclusion

# References and acknowledgements

## INVESTIGATION

**Investigation number**

**Title**

**Introduction**

**Aim**

**Hypothesis**

| Name: | Your signature: | Date: |
|---|---|---|
| Partner: | Partner's signature: | Teacher's signature: |

## Methodology

## Materials

## Method

**Results**

## Discussion

Name:

Partner:

Your signature:

Partner's signature:

Date:

Teacher's signature:

9780170452625

## Conclusion

_____

_____

_____

_____

_____

_____

_____

_____

_____

_____

_____

_____

## References and acknowledgements

_____

_____

_____

_____

_____

_____

_____

_____

_____

_____

_____

_____

_____

Name: _____ Your signature: _____ Date: _____

Partner: _____ Partner's signature: _____ Teacher's signature: _____

## INVESTIGATION

**Investigation number**

**Title**

**Introduction**

**Aim**

**Hypothesis**

## Methodology

## Materials

## Method

| Name: | Your signature: | Date: |
| Partner: | Partner's signature: | Teacher's signature: |

**Results**

# Discussion

| Name: | Your signature: | Date: |
|---|---|---|
| Partner: | Partner's signature: | Teacher's signature: |

## Conclusion

## References and acknowledgements

# Key terms

**5**

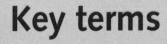

Here is a list of the essential key terms that you will need to be familiar with when performing an investigation. Complete the key term definitions so you become comfortable using and applying them in your investigations.

## Science investigation methodologies

| Key term | Write your own definition so you understand the meaning of each term |
|---|---|
| Case study | |
| Classification and identification | |
| Controlled experiment | |
| Fieldwork | |
| Literature review | |
| Modelling | |
| Product, process or system development | |
| Simulation | |

Name:                    Your signature:                    Date:

Partner:                    Partner's signature:                    Teacher's signature:

## Data and measurement

| Key term | Write your own definition so you understand the meaning of each term |
|---|---|
| Accuracy | |
| Precision | |
| Repeatability | |
| Reproducibility | |
| True value | |
| Validity | |

## Errors, uncertainty and outliers

| Key term | Write your own definition so you understand the meaning of each term |
|---|---|
| Personal errors | |
| Random errors | |
| Systematic errors | |
| Uncertainty | |
| Outliers | |

## Bioethics

| Key term | Write your own definition so you understand the meaning of each term |
|---|---|
| **Approaches to bioethics** | |
| Consequences-based approach | |
| Duty- and/or rule-based approach | |
| Virtues-based approach | |
| **Ethical principles** | |
| Integrity | |
| Justice | |
| Beneficence | |
| Non-maleficence | |
| Respect | |

Name:  Your signature:  Date:

Partner:  Partner's signature:  Teacher's signature:

Name: _____   Your signature: _____   Date: _____

Partner: _____   Partner's signature: _____   Teacher's signature: _____   9780170452625

Name:                    Your signature:                    Date:

Partner:                 Partner's signature:               Teacher's signature:

Name: _____ Your signature: _____ Date: _____

Partner: _____ Partner's signature: _____ Teacher's signature: _____ 9780170452625

Name: _____ Your signature: _____ Date: _____

Partner: _____ Partner's signature: _____ Teacher's signature: _____ 9780170452625

Name: _____   Your signature: _____   Date: _____

Partner: _____   Partner's signature: _____   Teacher's signature: _____

Name: _____          Your signature: _____          Date: _____

Partner: _____          Partner's signature: _____          Teacher's signature: _____          9780170452625

Name: _____   Your signature: _____   Date: _____

Partner: _____   Partner's signature: _____   Teacher's signature: _____

Name: _____  Your signature: _____  Date: _____

Partner: _____  Partner's signature: _____  Teacher's signature: _____  9780170452625

| Name: | Your signature: | Date: |
| --- | --- | --- |
| Partner: | Partner's signature: | Teacher's signature: |

Name: _____  Your signature: _____  Date: _____

Partner: _____  Partner's signature: _____  Teacher's signature: _____  9780170452625

Name: _____    Your signature: _____    Date: _____

Partner: _____    Partner's signature: _____    Teacher's signature: _____

| Name: | Your signature: | Date: |
|---|---|---|
| Partner: | Partner's signature: | Teacher's signature: |

Name: _____  Your signature: _____  Date: _____

Partner: _____  Partner's signature: _____  Teacher's signature: _____  9780170452625

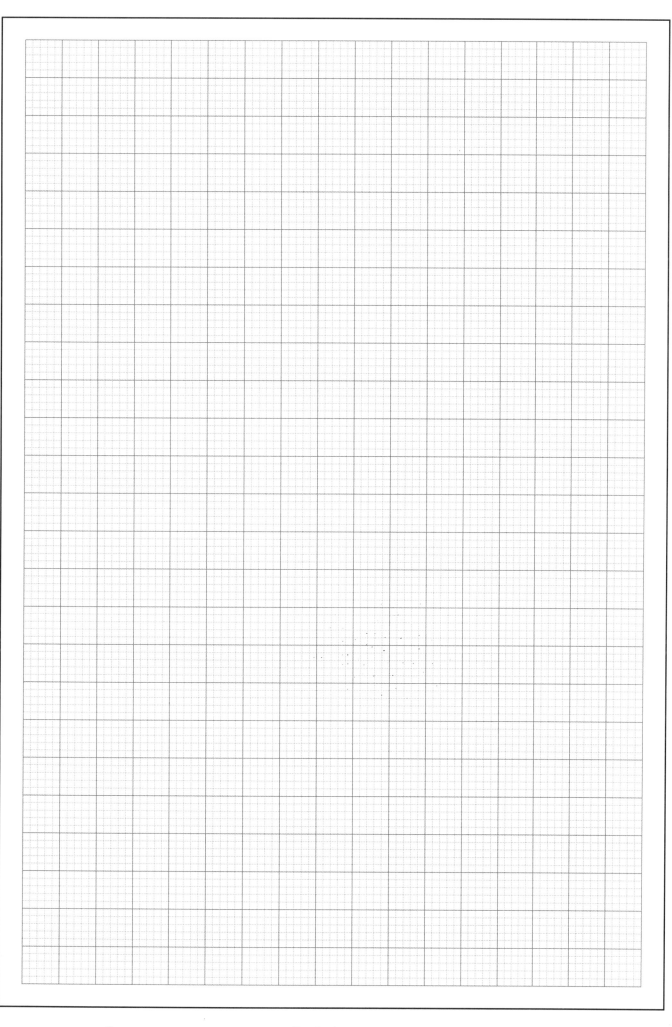

| Name: | Your signature: | Date: |
| Partner: | Partner's signature: | Teacher's signature: |